GEMS
OF SOUTHERN AFRICA
A FIRST FIELD GUIDE

Beryl (aquamarine), page 20

Quartz, page 42

BRUCE CAIRNCROSS

Contents

Introduction — 3
  What are gemstones? — 3
  How gemstones are formed — 4
  Identifying gemstones — 5
  Naming gemstones — 6
  Physical characteristics — 7
  Mining of gemstones — 9
  Artificial gemstones — 11
  Learn more about gemstones — 12
  Crystal systems — 13
  How to use this book — 16
  Birthstones — 17
Species accounts — 20
Glossary — 54
Further reading — 56
Index and Checklist — 57

First edition published in 2001 by
Struik Publishers (a division of New Holland Publishing, South Africa (Pty) Ltd)
80 McKenzie Street
Cape Town 8001
www.struik.co.za

New Holland Publishing is a member of the Johnnic Publishing Group.

10 9 8 7 6 5 4 3 2 1

Copyright © 2001 in text:
  Bruce Cairncross
Copyright © 2001 in photographs:
  Bruce Cairncross.
Copyright © 2001 maps and illustrations:
  Struik Publishers
Copyright © 2001 in published edition:
  Struik Publishers

ISBN 1 86872 599 5

Managing Editor: Helen de Villiers
Editors: Mehita Iqani,
  Katharina von Gerhardt
Designer: Bridgitte Chemaly

Reproduction: Scan Shop
Printed by: CTP Book Printers (Pty) Ltd

All rights reserved. No part of this publication may be reproduced, stored in a retrieval system or transmitted, in any form or by any means, electronic, mechanical, photocopying, recording or otherwise, without the prior written permission of the copyright holders.

Introduction

What are gemstones?

Most gemstones are types of naturally occurring mineral or rock. Some gemstones, such as pearl and coral, however, occur organically, which means that they are produced by living matter (plants or animals). In order for a substance to be defined as a gem, it must fulfil three main criteria. Firstly, it must be beautiful. This usually means that it has an attractive colour and relates to the way it reflects light. It also has to be durable and rare.

Gemstones are usually cut and polished, and are used for personal adornment or ornamental purposes.

In the past the terms 'precious' and 'semi-precious' were used to classify gems. The subdivision led to confusion in terminology and to the view that 'semi-precious' stones were inferior. For these reasons, the subdivision is now discouraged and all substances that are rare, durable and beautiful enough are called gemstones.

'Precious' stones (diamond, ruby, emerald, sapphire, pearl and sometimes opal and alexandrite) are rare and considered beautiful.

Emerald

'Semi-precious' stones (aquamarine, topaz, tourmaline and amethyst) are more common and less expensive.

Aquamarine

The rubicon pegmatite mine, Namibia

Southern Africa contains a wealth of diverse rocks and minerals. Some of the world's oldest rocks, which formed over 3 000 million years ago, are found here. Gemstones occur in three types of rock systems: igneous^G, metamorphic^G and sedimentary^G. Most gemstones can only be found in very specific rock types, which is what makes them so hard to find. A knowledge of rock types increases the chances of finding gemstone deposits. Some gemstones, such as diamond, carnelian, agate, chalcedony and jasper, can also occur in alluvial^G sediments of rivers, such as the Vaal and Gariep (Orange) rivers in South Africa.

How gemstones are formed

Gemstones are formed within the Earth's crust in the presence of particular physical and chemical conditions. Most minerals and gemstones are composed of compounds containing more than one element. Under the very great pressure and high temperature conditions that exist in the Earth's crust, elements can come together to form crystals of minerals and gemstones. This process can be fairly quick, but usually takes a very long time (thousands of years) to occur.

A crystal can be described as being a symmetrical solid with flat (or planar) faces. The particular shape of a crystal reflects the internal arrangement of the atoms, which bind together in a regular three-dimensional framework. Some minerals readily form well-shaped crystals, while others form crystals only rarely. The same mineral may form a variety of different crystal shapes depending on the particular nature of the forces that act on the chemical elements. So, for instance, carbon can produce either diamonds, under immense pressure – or graphite, in less extreme circumstances.

Some minerals react chemically with other minerals, or can be affected by external conditions. When this happens, the original mineral gets replaced by a second mineral, forming what is called a pseudomorph[G] (or false form). This means that the newly formed mineral closely resembles the original, but has a very different chemical composition.

Identifying gemstones

Some basic physical observations, described in 'Physical characteristics' on p. 7, can help to identify gemstones and minerals. Remember that all non-organic gemstones are minerals. Accurately identifying a gemstone can be a frustrating and difficult task for the keen amateur. In many instances it

Topaz (left) and diamond (right)

may be a good idea to enlist the help of a professional gemologist or jeweller to help identify rough, or cut and polished, gemstones. They can use specialized optical, physical and chemical tests that are either beyond the scope of an amateur collector, or that require specific equipment.

Emerald (left) and goshenite (right). Although these two stones look very different, they actually belong to the same species – beryl.

In the same way that one motor car manufacturer produces several different makes of motor car under one company name, different varieties of gemstone may be produced by a single species[G]. Ruby and sapphire are good examples of this phenomenon: they are both varieties of corundum.

> **Did you know?**
> The gemstone with the most varieties is quartz, which includes agate, carnelian, citrine, tiger's eye, amethyst and aventurine, among others.

Naming gemstones

Gemstones are named and classified according to international convention. A distinction should be made between formally recognized names, which have to be approved by the International Mineralogical Association (IMA), and informal names used by the gemstone trade and by amateurs. This can cause confusion and should be avoided, as different names are often used for the same gemstone. Common names, such as 'wonderstone', referring to pyrophyllite, are widely used and recognized, and for this reason they are included in this field guide.

Physical characteristics

The physical properties of a gemstone are described by various characteristic terms, listed below. The best way to recognize a gemstone is to use as many criteria as possible, because one physical attribute alone is insufficient for accurate identification.

Colours: Some gemstones have only one colour and can be identified by it.

Ruby is always red.

Other types of gemstone such as sapphire, however, can have several colours. This criterion is, therefore, not fully conclusive.

Streak: Some minerals produce a distinctively coloured streak when scratched on an unglazed white porcelain tile. Most gemstones, however, produce a colourless streak.

Lustre: Lustre ranges from metallic to vitreous (like broken glass), resinous (like resin), pearly, silky and adamantine (the lustre of a diamond). After being intensely polished, all gemstones have an attractive lustre.

Heliodor (beryl) has a vitreous lustre.

Transparency: Gemstones can be transparent (see-through), translucent (let light through) or opaque (do not let light through).

Form: Uncut gemstones can show clearly defined crystal faces, or they can be amorphous[G] and have no visible external crystal form.

Habit: This refers to the specific natural shapes of gemstones or minerals, and describes how they look. These include:

fibrous:	fine and hair-like
acicular:	needle-like
bladed:	like a knife blade
tabular:	broad and flat
prismatic:	elongated in one direction
stellate:	radiating from central point
botryoidal:	resembling a bunch of grapes

Tenacity: This refers to how a gemstone or mineral reacts to being crushed or broken. For example, it can be brittle and shatter; tenacious and resist breaking; flexible and bend a little before breaking or malleable and be shaped, bent and stretched without breaking. Most gems are brittle.

Cleavage: This refers to the way gemstones or minerals split along sets of defined planes. The cleavage pattern relates strongly to the internal atomic structure of the gemstone.

Specific gravity (SG): This relates to density – the atomic weight of constituent elements and the way they are packed together. The SG is the ratio of a gemstone or mineral's weight to the weight of an equal volume of

Almandine garnet (left; SG = 4,1-4,3) is heavier than common opal (right; SG = 1,9-2,4).

water. Determining the SG of a gemstone can be very useful in identifying it. The greater the SG, the heavier the species^G.

Hardness: This is a good measure of the resistance of minerals to abrasion, and a good indicator of identity. The most common method of checking how hard gems are is Mohs' Hardness Scale. It is a simple and effective technique that compares the hardness of minerals by scratching them against each other. The scale goes from 1 (softest) to 10 (hardest). Each mineral can scratch those that score lower than it on the scale. Corundum can scratch all minerals except diamond. In practice, some everyday objects can be used to test the hardness of minerals. A fingernail is about 2,5 hard and can scratch talc and gypsum. Common window glass has a hardness of about 5,5 and a common penknife blade a hardness of about 6. Most gemstones are, by definition, relatively hard.

Mohs' Hardness Scale

1 = talc
2 = gypsum
3 = calcite
4 = fluorite
5 = apatite
6 = feldspar
7 = quartz
8 = topaz
9 = corundum
10 = diamond

Amethyst (quartz) = 7

Mining of gemstones

Collecting gemstones ranges from the simplest methods, such as panning for stones in a riverbed, to vastly expensive operations that are dependent on modern technology.

In southern Africa, gemstones are mined in many different ways, from the huge opencast and expansive diamond mines in South Africa, to very informal, small-scale digging for topaz gemstones in Namibia.

Many small-scale mining operations are plagued by the problem of theft of gemstones.

Tafelkop, close to Brandberg, Namibia, is a source of beautiful quartz and amethyst.

Gemstone mines that extract tourmaline, topaz and emeralds, for example, require individual miners to hand-sort the gems from the rock, and strict security is needed to safeguard against theft.

The traditional method of panning is still practised by many communities around the world such as in Sri Lanka, where sapphires are panned from rivers. Panning involves sifting through deposits of sand and gravel. The swirling action of water is used to wash away lighter material, leaving behind only the heavier material, including any gemstones.

Some gemstones, such as diamonds, can be found on the Earth's surface in alluvial[G] deposits, but are more often embedded in volcanic rock such as kimberlite[G]. Modern mining processes are required to unearth them. Modern diamond mines are constructed on a vast scale, penetrating deep into the Earth's crust, and requiring over 250 tonnes of rock to be blasted for every finished diamond carat (0,2 g).

Some diamonds are found on the ocean floor. In this case, offshore trawling is conducted by large ships that pump up gravel from the seabed. Diamonds are trawled in this way off the coast of Namibia.

Many of the gemstone mines in South Africa, Namibia and Zimbabwe extract gems from pegmatite[G] rock. Pegmatites[G] are always relatively small geological deposits, usually not more than a few hundred metres long and a few tens of metres wide. Almost all of the small mines that exploit these deposits consist of opencast pits or quarries. The surrounding rock is first exposed, then the zones in the pegmatite^G that contain gemstones are identified. These can be very narrow, and can take the form of pockets or hollows scattered randomly in the deposit. Mining techniques involve either digging by hand or with compressor-driven tools such as jackhammers, because if explosives are used, the force from the blast would shatter any gemstones, rendering them useless.

De Beers diamond mine, Kimberley

Did you know?
Over 5 000 years ago in Ancient Egypt, the Pharaohs employed thousands of slaves to operate the first commercial gemstone mines.

Artificial gemstones
The fact that rare gemstones are so sought after has led to the creation of industries that can manufacture artificial gemstones on demand. There are three basic methods of copying gemstones:

Imitation stones look like the real thing, but have a different composition. An example of an imitation stone is cubic zirconia; a gemstone created from this material closely resembles a diamond, although its chemical composition is quite different.

Synthetic stones are chemically almost identical to natural gems but are created in laboratories. Synthetic diamonds with an identical crystalline structure to gems found in the Earth's crust have been created in laboratories. Known as industrial diamonds, these are created more for their use in industry than for their visual splendour. However, recent technological developments have allowed for the production of synthetic diamonds of several carats that rival some of the natural, high-quality gemstones.

Composite stones consist of several elements cemented together. They often consist of a sliver of genuine gemstone set on a base of, for instance, glass. The general effect is pleasing, although the value of the item is small.

Artificial gemstones that are grown in laboratories nowadays are so well made that it takes a true expert to closely examine these gems with a microscope to be able to tell them apart from their natural counterparts.

Did you know?

In 1891, the French scientist Auguste Verneuil perfected a technique for producing synthetic gems. He sifted powdered crystals into a flame, melting them onto a holder. Once the melted crystal had been removed from the heat, it cooled and formed a solid crystal.

Learn more about gemstones

- Visit your local museum, view its collection and talk to the curator.
- Visit the Geology Departments at Universities and Technikons, view their collections and talk to the staff.
- Join your local club. All major cities in South Africa have amateur clubs, which can be contacted via the Federation of South African Gem and Mineral Societies (FOSAGAMS).
- Contact mining companies.
- For special queries contact the

Geological Societies of South Africa, Namibia, Botswana, Swaziland, Zimbabwe or Mozambique.
- Read books about gemstones and minerals and watch television documentaries.
- Visit websites of museums, gem dealers and collectors with home pages.
- Subscribe to magazines: *Gems & Gemology* published by the Gemological Institute of America (GIA), or the *Mineralogical Record* published by the Mineralogical Record INC in Tucson, Arizona.
- Collect rocks and minerals.
- Be careful, avoid injury and never trespass on private property. Try to find an experienced collector to accompany you.

Crystal systems

All known minerals can be classified into one of six crystal systems. All members of a crystal system, whether gemstones or other minerals, have relatively common interior traits, even though they have very complex exterior crystal face patterns.

A basic crystal shape may be common to a crystal system, but in nature can easily be modified by other complex forms. Thus identification of gemstones based on crystal shape can be difficult.

Topaz (top) – orthorhombic crystal system. Garnet (andradite) (bottom) – cubic crystal system

CUBIC CRYSTAL SYSTEM:		$+a_3$ all angles 90° $-a_1$ $-a_2$ —— $+a_2$ $+a_1$ $-a_3$
TETRAGONAL CRYSTAL SYSTEM:		$+c$ all angles 90° $-a_1$ $-a_2$ —— $+a_2$ $+a_1$ $-c$
ORTHORHOMBIC CRYSTAL SYSTEM:		$+c$ all angles 90° $-a$ $-b$ —— $+b$ $+a$ $-c$
MONOCLINIC CRYSTAL SYSTEM:		$+c$ $>90°$ $-a$ $-b$ —— $+b$ $+a$ $<90°$ $-c$
TRICLINIC CRYSTAL SYSTEM:		$+c$ all angles ≠ 90° $-a$ $-b$ —— $+b$ $+a$ $-c$
HEXAGONAL CRYSTAL SYSTEM:		$+c$ 90° $-a_3$ $-a_1$ $+a_2$ $-a_2$ $+a_1$ $+a_3$ $-c$

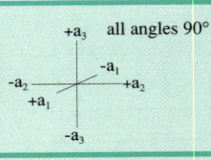

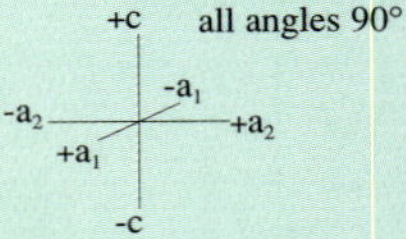

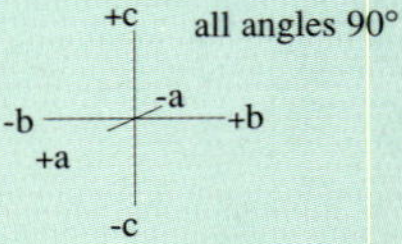

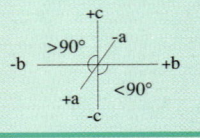

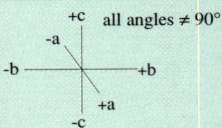

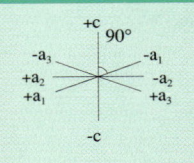

- Three imaginary axes of the same length ($a_1 = a_2 = a_3$).
- All at right angles to one another.
- Basic shape is a perfect cube.
- Common forms are the cube, octahedron[G] and dodecahedron[G].
- Garnet, diamond and sodalite form in this system.

- Three imaginary axes with two horizontal axes of equal length and a shorter or longer vertical axis ($a_1 = a_2 \neq c$).
- All at right angles to one another.
- Crystals have a shoebox shape.
- Zircon and rutile form in this system.

- Three imaginary axes of different lengths ($a \neq b \neq c$).
- At right angles to each other.
- Crystals have a matchbox shape.
- Topaz, chrysoberyl, sphene and iolite form in this system.

- Three imaginary axes of different lengths ($a \neq b \neq c$),
- Two at right angles to one another, the third inclined at an angle.
- Jade, dumortierite, sphene and some feldspars form in this system.

- Three imaginary axes of different lengths ($a \neq b \neq c$).
- Inclined to one another.
- Euclase crystallizes in this system.

- Four imaginary axes: three lateral axes of equal length intersecting at 120°. ($a_1 = a_2 = a_3 \neq c$) and a shorter or longer vertical axis
- At right angles to the other three.
- The basic shape is almost like a six-sided pipe, elongate with six crystal faces.
- The trigonal system has the same crystallographic relationships as the hexagonal system, but has different symmetry relationships.
- Quartz, beryl, corundum and tourmaline crystallize in the trigonal system.

How to use this book

In this book, gemstones have been classified alphabetically according to species[G]. For this reason, some better-known gemstones may not have their own section. The gem you are looking for may be mentioned in the subheading 'Varieties', and listed in the index on page 57. Each species[G] entry begins with a general introduction, followed by a list of varieties that make up the species[G]. A more detailed account follows, split up into several headings, listed below, all of which should help you in identifying gems:

> **Did you know?**
> The 'Did you know' boxes highlight relevant, exciting and interesting information.

Varieties and colours: 'Varieties' refers to all those gemstones belonging to this species[G]. Although some gemstones may only have one colour, others may have several.

Crystal system and mineral group: Indicates to which one of the six crystal systems the gemstone belongs (*see* pp. 13, 14, 15 for further information). The mineral group refers to different minerals categorized into groups on the basis of similar chemical composition.

Composition: The chemical elements of which it consists.

Hardness: Refer to Mohs' Hardness Scale (*see* p. 9 for explanation). Gives a good indication of identity of gemstone. Hardness (H) is indicated in a box at the top, right-hand corner of each species[G] account.

Specific gravity: Refers to the density of the gem. This (SG) is also noted in a box at the top, right-hand corner of each species[G] account.

Lustre: The surface reflection of a natural, unpolished gem.

Amethyst (quartz) SA: 5,2 cm

Caption: Includes the country of origin of gemstone and size (mm or cm of the longest dimension) or weight (carats).

Abbreviations of countries of origin:
Moz = Mozambique
Nam = Namibia
SA = South Africa
Zam = Zambia
Zim = Zimbabwe

Identification points: Highlight the telling characteristics specific to that gem.

Size: Range of its size.

Status: This indicates whether the gem is easy to find (common), or if it is rare.

Economic value: Indicates what it is used for, e.g. for carvings. lapidary material, etc. Rareness increases economic value.

Distribution: Refers to the areas in southern Africa where these mineral deposits occur, sometimes highlighting extraordinary finds.

Words followed by an uppercase[G] are explained in the glossary on page 54.

> **Did you know?**
> Gemstones have been associated with different months of the year since the 1st century AD.

Birthstones
Wearing birthstones for luck became a popular custom in the eighteenth century in Poland. Overleaf, the gemstones associated with each month of the year are depicted.

Birthstones

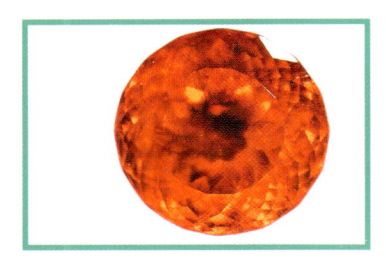

January Garnet

February Amethyst

March Aquamarine

April Diamond

May Emerald

June Pearl

July Ruby

October Opal

August Peridot

November Topaz

September Sapphire

December Turquoise

Beryl

H 7,5-8

SG 2,6-2,9

Beryl always forms six-sided crystals, and is usually pale green. The transparent varieties of beryl are very popular. Emerald is the most famous and valuable variety of beryl. Found in pegmatites[G] and quartz veins[G] associated with metamorphic[G] rocks.

Varieties and colours:
Aquamarine (blue), emerald (dark-green), goshenite (colourless), heliodor (yellow), and morganite (pink). Common beryl can be colourless, opaque white and pale green.

> **Did you know?**
> A beryl crystal weighing an amazing 106 kg was discovered in Swaziland.

Crystal system and mineral group: Hexagonal silicate[G].

Composition: Beryllium-aluminium silicate[G].

Lustre: Vitreous.

Identification points: Hexagonal crystals, hardness.

Size: Few cm to tens of cm.

Status: Rare (emerald, morganite, aquamarine); common (common beryl).

Economic value: Gemstone (emerald, aquamarine, heliodor, and morganite).

Beryl (goshenite). Nam: 3,7 cm

Beryl (heliodor). Nam: 13,6 carats

Distribution: Aquamarine deposits occur in the Northern Cape, the Northern Province and Mpumalanga in South Africa; Hurungwe and Mwami in Zimbabwe and Karibib and the Namib in Namibia. World-class emeralds are mined in Sandawana, Zimbabwe, and a deposit occurs in Gravelotte, South Africa. Heliodor occurs in the Northern Province of South Africa and close to Rössing Siding in Namibia. Morganite occurs in Namaqualand and Mpumalanga, South Africa and Etiro, Namibia – where extraordinary 3 m long crystals have been found. Common beryl occurs widely in South Africa and Zimbabwe.

Beryl (emerald). Zim: 4 carats

Beryl (aquamarine). Zim

Chrysoberyl

H 8,5

SG 3,75

Occurs in biotite[G] schists[G] as V-shaped, repeatedly twinned[G] crystals.

Did you know?
Alexandrite is a variety of chrysoberyl. It has the very strange visual property of changing colour when looked at under different light sources.

Varieties and colours: Grey, yellow, yellow-green, olive-green and dark green. Alexandrite (green in daylight and red in artificial light). Cat's-eye chrysoberyl is yellowish green in colour and displays asterism[G].

Chrysoberyl. Zim: 12 mm

Crystal system and mineral group: Orthorhombic oxide[G].

Composition: Beryllium-aluminium oxide[G].

Lustre: Vitreous.

Identification points: Twinned[G] sixling[G] crystals, change in colour under different light.

Size: Few mm to several cm.

Status: Very rare.

Economic value: Rare gemstone.

Distribution: Chrysoberyl occurs in the Northern Cape, Mpumalanga and Northern Province, South Africa. Golden chrysoberyl is found in Mwami and Somabula in Zimbabwe, which also boasts deposits of alexandrite at the Novello claims, east of Masvingo and cat's-eye chrysoberyl in Karoi.

Chrysocolla

H 2,4

SG 2,01-2,4

Found as a secondary oxide[G] mineral in the oxidized[G] zone of copper deposits.

Colours: Blue, blue-green.

Crystal system and mineral group: Monoclinic silicate[G].

Composition: Hydrous copper silicate[G].

Lustre: Earthy.

Identification points: Distinctive sky-blue colour, softness.

Size: Forms as huge lumps[G], not as microcrystals. The latter are visible to the naked eye.

Status: Rare.

Economic value: Lapidary[G] material.

Distribution: In South Africa, chrysocolla is found in small amounts with copper deposits, such as those at Messina and Okiep. In Namibia it is mined with dioptase in Kaokoland. In Zimbabwe it is found at the Elephant Mine, Devure Ranch.

Did you know?
Chrysocolla does not form as well-shaped crystals, but as massive lumps[G].

Chrysocolla. Nam: 6 cm polished egg

Cordierite

H 7-7,5

SG 2,53-2,78

Cordierite is an exclusively metamorphic[G] mineral, occurring only in metamorphic[G] gneiss[G] and schist[G]. The gemstone exhibits strong pleochroism[G].

> **Did you know?**
> As a transparent gemstone, cordierite is referred to as iolite or water-sapphire because of its violet-blue colour.

Varieties and colours: Iolite/water-sapphire (blue, violet, grey, brown, and yellow-green).

Crystal system and mineral group: Orthorhombic silicate[G].

Composition: Aluminium-magnesium silicate[G].

Lustre: Vitreous.

Identification points: Gem variety displays dichroism[G].

Size: Few mm to several cm. Sometimes it may form as massive lumps[G].

Status: Rare.

Economic value: Gemstone (iolite).

Distribution: In South Africa, the Hout Bay area in the Western Cape, Namaqualand in the Northern Cape and the Marble Delta, KwaZulu-Natal. In Namibia, alluvial[G] iolite is found between Walvis Bay and Rooibank and in the Namib. Gem-quality blue iolite occurs in southern Zimbabwe and in the Karoi district.

Cordierite (iolite rough). Zim: 3 cm pieces

Corundum

H 9

SG 4-4,1

Corundum occurs mainly in metamorphic^G rocks and is the second hardest gemstone after diamond.

Varieties and colours: Ruby (red), sapphire (colourless, grey, brown, orange, yellow, green, purple, or blue). Star sapphires display asterism^G, six-armed, star-like reflections, when the crystal is cut *en cabochon*^{G}.

Corundum (sapphire). Zim: 4,35 carats

Crystal system and mineral group: Trigonal oxide^G.

Composition: Aluminium oxide^G

Lustre: Vitreous, adamantine.

Identification points: Hexagonal, barrel-shaped crystals, hardness.

Size: Few mm to tens of cm.

Status: Rare (ruby, sapphire) and common (corundum).

Economic value: Gemstone (ruby, sapphire).

Distribution: In South Africa, corundum is found in the Northern Province, Northern Cape (Namaqualand), the Lowveld, and KwaZulu-Natal. Rubies and sapphires are rarely found. In southern Zimbabwe ruby and sapphire are mined from the Somabula conglomerate. Sapphire- blue corundum occurs in alluvial^G deposits in north-east Botswana, where gemstones of up to 3 cm have been discovered. Eluvial^G deposits in Swaziland in the Hlatikulu-Goedgegun area.

Dumortierite

H 8,5

SG 3,41-3,71

Dumortierite is found as massive lumps[G] (not crystals) and has a very distinctive violet-lilac colour.

Colours: Blue, violet-blue, pink, and brown.

Crystal system and mineral group: Orthorhombic silicate[G].

Composition: Aluminium borosilicate[G].

Lustre: Vitreous to dull.

Identification points: Colour.

Size: Forms as massive lumps[G].

Status: Rare.

Economic value: Lapidary[G] material.

Distribution: Found in gneisses[G] in Namaqualand, South Africa; north of the Erongo Mountains. In Namibia; in the Tete area in Mozambique and in the Zimbabwean lowveld.

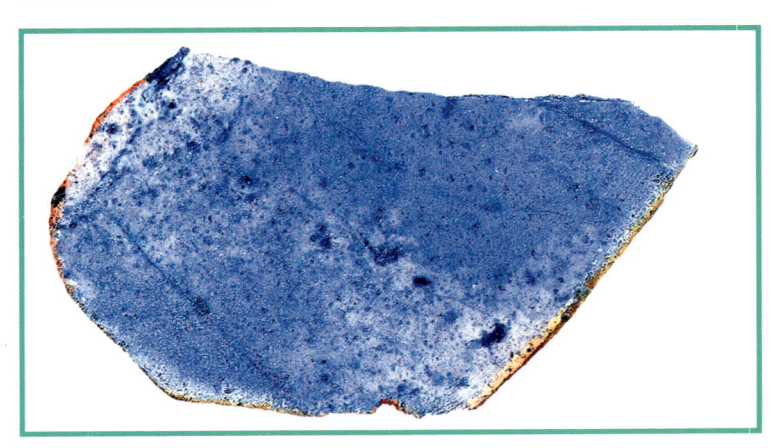

Dumortierite. Moz: 10 cm

Euclase

H 7,5

SG 3,05-3,10

Euclase is a very rare gemstone, and is only found in north-eastern Zimbabwe. It is usually an attractive blue colour, and forms by the chemical replacement of beryl crystal, thus forming a pseudomorph[G].

Colours: Blue, pale blue, pale green, or colourless.

Crystal system and mineral group: Monoclinic silicate[G].

Composition: Beryllium silicate[G].

Lustre: Vitreous.

Identification points: Appealing vibrant blue colour.

Size: Few mm to few cm.

Status: Very rare.

Economic value: Rare gemstone.

Distribution: Euclase is found only in pegmatite[G] rock in the Mwami area in Zimbabwe.

Euclase. Zim: 5,5 cm

Feldspar group

H 6-6,5

SG 2,5-2,6

The feldspars are a group of rock-forming minerals found in igneous^G and metamorphic^G rocks. Sometimes feldspars form as large crystals with various colours. They can also be transparent, which makes them popular gemstones.

Varieties and colours: Albite or adularia, which is known as moonstone (translucent, milky); green microcline, which is known as amazonite (green); unakite, which is a mixture of pink orthoclase, green epidote and quartz (pinkish green); labradorite (blue-green).

Crystal system and mineral group: Monoclinic/triclinic/orthorhombic silicates^G.

Composition: Potassium-aluminium silicate^G.

Lustre: Vitreous, pearly.

Identification points: Crystals have good cleavage.

Size: Few mm to tens of cm.

Feldspar (unakite). SA: 6 cm polished egg

Status: Relatively common.

Economic value: Lapidary^G material.

Distribution: Amazonite occurs in the Northern Province, and Northern Cape, South Africa; in the Maltahöhe district and south of Otjiwarongo in Namibia; in Mwami, Hurungwe and Pfungwe in Zimbabwe; and close to the Zimbabwe border in Mozambique. Unakite occurs in the Northern Cape in South Africa and in the Beitbridge, Gweru and Bulawayo areas in Zimbabwe.

Garnet group

In medieval times garnets were thought to have certain powers and magical properties. There are several quite different varieties of garnet exhibiting an array of different colours, but one common feature is the crystal system in which they form – the cubic crystal system. It forms in metamorphic[G] rocks such as schists[G] and in pegmatites[G] and igneous[G] rocks such as kimberlite[G].

> **Did you know?**
> It has been suggested that the name garnet stems from the Latin word for pommegranate; a fruit which has red, grain-like seeds.

Garnet (andradite). SA: 8 mm crystals

Almandine garnet

H 7-7,5

SG 4,1-4,3

Colours: Dark red, red-brown, violet-red, and brown-black.

Crystal system and mineral group: Cubic silicate^G.

Composition: Iron-magnesium-aluminium silicate^G.

Lustre: Vitreous, resinous.

Garnet (almandine). SA: 4,2 cm

> **Did you know?**
> Almandine is one of the most widespread varieties of garnet, and often occurs as a large, perfect crystal.

Identification points: Hardness. Dodecahedral^G crystals with many crystal faces.

Size: From a few mm to several cm.

Status: Common.

Economic value: Gemstone.

Distribution: In the Northern Cape's Namaqualand and the Soutpansberg-Limpopo region of South Africa. In schists^G in the Karibib-Swakopmund districts of Namibia. The Mutoko and Beitbridge areas and Hurungwe in Zimbabwe. High-quality gems stem from Tuli in Botswana, and there are many deposits in the ancient Basement rocks in eastern Botswana.

Andradite garnet

H 6,5-7

SG 3,7-4,1

Varieties and colours: Melanite (black); demantoid (coloured green by chrome); topazolite (yellow). Common andradite characteristically dark brown to black, can also occur as yellow-green, brown and red-brown.

Crystal system and mineral group: Cubic silicate[G].

Composition: Calcium-iron silicate[G].

Lustre: Vitreous, resinous.

Identification points: Dodecahedral[G]-shaped crystals; common andradite is usually dark brown to black.

Size: Few mm to few cm.

Status: Rare (demantoid); common (other andradite varieties).

Economic value: Gemstone (demantoid).

Distribution: Some 10 carat stones (ranging in colour from green to yellow-green to blue-green) have been found on a farm called Tubussis in central Namibia. Demantoid variety found in the Chimanda Communal Lands in Zimbabwe. In South Africa restricted to the Kalahari manganese field.

Garnet (andradite). SA: 3 cm

Grossular garnet

H 7-7,5

SG 3,5-3,8

Colours: Green, orange, red, pink, yellow.

Crystal system and mineral group: Cubic silicate^G.

Composition: Calcium-aluminium silicate^G.

Lustre: Vitreous.

Identification points: Common green or orange colours.

Size: Few mm to few cm.

Status: Rare.

Economic value: Used as a gemstone when transparent, or for lapidary^G purposes if opaque.

Distribution: In South Africa, gem-quality red-orange grossular occurs in southern KwaZulu-Natal; other deposits occur in the Northern Province. A very attractive, massive hydro-grossular^G garnet occurs north of Johannesburg in the bushveld region. It is called 'Transvaal jade' due to its translucent green, pink and red colours.

> **Did you know?**
> Grossular gets its name from the botanical term for gooseberry (*Ribes grossularia*), which it resembles in colour and shape.

Note: 'Transvaal jade' is often grouped among the jade group. However, it is a form of garnet, found in Gauteng and the North-West Province, South Africa.

Garnet (Transvaal jade). SA

Pyrope garnet (Cape ruby)

H 7-7,5

SG 3,8-4,3

Garnet – pyrope (rhodolite). Zim: 7 carats

Varieties and colours: Rhodolite (red-violet). Can also be red, pink-red, orange-red, and purple-red.

Crystal system and mineral group: Cubic silicate^G.

Composition: Magnesium-aluminium silicate^G.

Lustre: Vitreous.

Identification points: Characteristic red colour, often transparent.

Size: Few mm to few cm.

Status: Rare.

Economic value: Rhodolite is a valuable gemstone when transparent.

> **Did you know?**
> The blood-red colour of the pyrope is due to its iron and chromium content.

Distribution: Found in the alluvial^G gravel of diamond diggings in the Free State, Northern Cape and North-West Province in South Africa and in colossus kimberlite^G pipe in the Mutoko and Beitbridge areas of Zimbabwe.

Garnet – pyrope (rhodolite). Zim: 10 mm pieces

Spessartine garnet

H 6,5-7,5

SG 3,4-3,6

Spessartine is a very valuable gem variety of garnet, and stones that come from northern Namibia are amongst the finest known. Gem-quality stones are rare.

Varieties and colours: Orange, red, yellow-brown, brown, may be transparent.

Crystal system and mineral group: Cubic silicate[G].

Composition: Manganese-aluminium silicate[G].

Lustre: Vitreous.

Identification points: Orange-red colour, dodecahedral[G] crystals.

Size: Few mm to few cm.

Status: Rare.

Distribution: Occasionally found in pegmatites[G] in the Northern Cape, South Africa. Gem-quality, orange spessartine garnets have been produced in the northern Kunune region of Namibia.

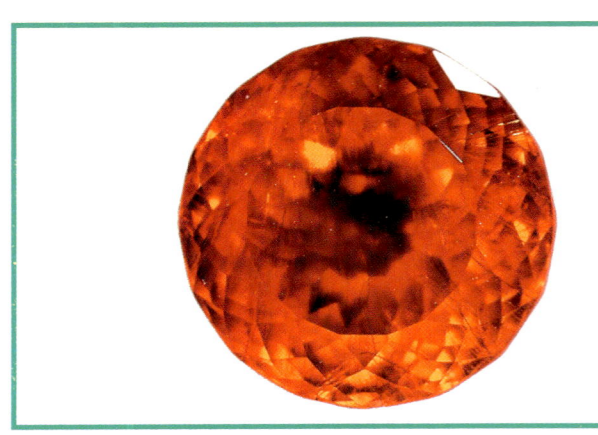

Garnet (spessartine). Nam: 12,6 carats

Jadeite and Nephrite (Jade)

H 6

SG 3,25

Jade has been used for centuries to carve beautiful ornaments. Jade is the common term for two minerals, jadeite and nephrite, both of which can be cut and polished.

Colours: Many shades of green or colourless.

Crystal system and mineral group:
Monoclinic silicate^G.

Composition: Sodium-aluminium silicate^G.

Lustre: Vitreous, greasy.

Identification points: Green, massive material.

Size: Massive, amorphous^G.

Status: Rare.

Economic value: Lapidary^G material, used for *objets d'art*^G.

Distribution: Dark to pale green nephrite jade occurs in serpentine in the Masvingo district, Zimbabwe.

Did you know?
Jade is an extremely tough gemstone that does not shatter easily.

Garnet (Transvaal jade). SA: 2,5 cm

37

Jeremejevite

H 7,5

SG 3,29

A very rare gemstone that is found in only a few places in the world. It forms very attractive blue crystals. Those discovered in Namibia are world famous amongst gem cutters due to their vibrantly blue colour, their formidable size and their transparency.

Colours: Colourless to pale blue (gem variety).

Crystal system and mineral group: Hexagonal borate[G].

Composition: Aluminium borate[G].

Lustre: Vitreous.

Identification points: Elongate, prismatic crystals.

Size: Few mm to a few cm.

Status: Extremely rare.

Economic value: Very rare gemstone.

Distribution: Found in granite from Mile 72 north of Swakopmund, Namibia. Also from the Erongo Mountains north of Karibib, where the finest crystals known were discovered early in 2001.

Jeremejevite. Nam: 1,9 cm

Malachite

H 3,5-4

SG 4,05

Malachite forms in both crystals and lumps[G] and the latter are used to produce beautiful ornaments, carvings and jewellery. The beautiful dark green colour is one of malachite's most popular properties.

Malachite. Zam: 20 cm

Colours: Green, black-green.

Crystal system and mineral group: Monoclinic carbonate[G].

Composition: Copper carbonate[G].

Lustre: Earthy, vitreous (crystals).

Identification points: Distinctive green colour, often with concentric whorls and patterns; reacts with acid.

Size: Crystals are small and lumps[G] are massive.

Status: Crystals are rare, while lumps[G] are common.

Economic value: Lapidary[G] material, used for carvings and *objets d'art*[G].

Distribution: Found associated with copper deposits in most southern African countries. Huge deposits in the Democratic Republic of Congo.

Opal (Common opal)

H 5,5-6,5

SG 1,99-2,25

The famous gem opal comes from Australia, and is sparkling and multicoloured. Gem opal does not occur in southern Africa, but common opal does.

Colours: Common opal is white, cream or brown, while precious opal can be colourless, blue, red, pink, brown or orange.

Crystal system and mineral group: Amorphous^G oxide^G.

Composition: Hydrated silicon dioxide.

Lustre: Vitreous, pearly, resinous.

Did you know?
Opal, as hydrated silica, contains about one-tenth water.

Identification points: Conchoidal^G fracture pattern; glassy appearance; light weight.

Size: Occurs as amorphous^G veins^G and lumps^G.

Status: Rare.

Economic value: Precious opal, gem and lapidary^G material. Common opal is sometimes cut and polished, but is much less valuable.

Opal. SA: 9,2 cm

Distribution: In South Africa common opal occurs in the Northern Cape, Northern Province, Mpumalanga, the Pilanesberg and the western regions of the Soutpansberg. It also occurs as a replacement mineral^G in fossilized wood in northern KwaZulu-Natal. In Zimbabwe, it is common as pale-green to white amorphous^G veins^G in many serpentinites and the Great Dyke.

Pyrophyllite (Wonderstone)

H 1-2

SG 2,65-2,9

Pyrophyllite is better known by its common name, wonderstone, which refers to its many different properties. When mined, it is a dull grey rock, but when carved and well polished, it shows different colours.

Did you know?
Pyrophyllite has such good thermal properties that wonderstone tiles were used on the nose cones of some NASA Challenger space ships.

Colours: Grey, brown, mottled.

Crystal system and mineral group: Monoclinic/triclinic silicate[G].

Composition: Aluminium silicate[G].

Lustre: Earthy.

Identification points: Extremely fine-grained and compact.

Size: Microscopic crystals.

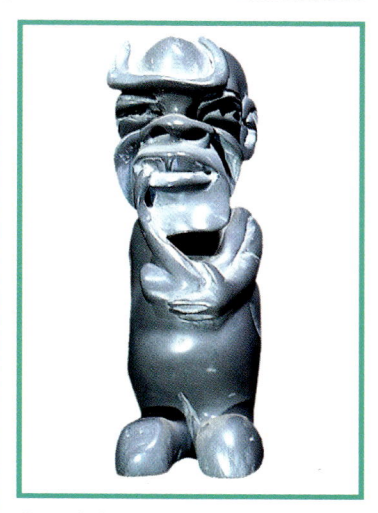

Pyrophyllite carving. SA: 12 cm

Status: Relatively rare.

Economic value: Used for carvings, and *objets d'art*[G].

Distribution: Fairly extensive deposits are found in South Africa from the Ottosdal area, which lies west of Klerksdorp and in Namibia, in the western region of the Brandberg Mountains.

Quartz

H 7

SG 2,66

Quartz usually forms distinctive six-sided crystals with sharply pointed terminations, but some varieties of quartz, such as agate, have no true crystal shape. Tiger's eye, a very popular gemstone, is actually made of quartz that has completely replaced asbestos, thereby forming a pseudomorph[G].

Varieties and colours: Agate: moss agate, picture agate, (both are multi-coloured), blue-lace agate (blue); amethyst (purple); aventurine (coloured green by mica); chalcedony (blue-grey); carnelian (orange-red); chrysoprase (coloured green by nickel); citrine (yellow); hyalite (colourless); jasper, including brecciated[G] jasper (coloured red by iron); milky quartz (white); rose quartz (pink); smoky quartz (grey-black); tiger's eye (brown or blue); mtorolite (coloured green by chrome); rock crystal quartz (colourless); rutilated quartz (colourless with orange-red, needle-like inclusions) and pietersite (mixture of blue and brown).

Did you know?
Quartz is the most common mineral on earth and is found in many rock types in many different shapes, sizes and colours. It occurs commonly in veins[G] that crosscut igneous[G] and metamorphic[G] rock.

Crystal system and mineral group: Trigonal oxide[G].

Composition: Silicon dioxide.

Lustre: Vitreous.

Quartz (tiger's eye). SA:16 cm (back left)

Quartz (blue lace agate). Nam: 20 cm

Identification points: Common as hexagonal crystals, many colour variations. No cleavage; conchoidal^G fracture pattern.

Size: Few mm to several m.

Status: Very common.

Economic value: Gemstone and lapidary^G material.

Distribution:
South Africa boasts deposits of amethyst, carnelian, chalcedony, jasper, tiger's eye, agate, aventurine and rose, smoky and milky quartz in various regions around the

Quartz (jasper). SA: 17 cm (left)

Quartz. SA: 10.9 cm

country. Deposits from the Jan Coetzee mine in the Northern Province and the Messina copper mines have yielded crystals of up to 1 m.

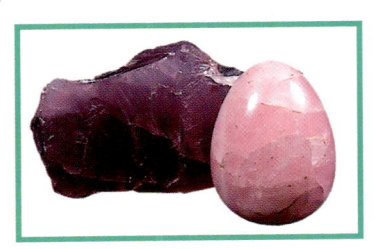

Quartz (rose quartz). Nam: 6 cm polished egg

Namibia is rich in many varieties of quartz, including agate, amethyst, carnelian, chalcedony, chrysoprase, citrine, jasper, hyalite, pietersite, rose and smoky quartz. Exciting deposits in Namibia are of high-quality, semi-transparent chalcedony north-east of Okahandja; jasper pebbles that occur on coastal beaches; amethyst-lined geodes from the Skeleton Coast; world-famous smoky quartz and amethyst from Tafelkop, west of the Brandberg and a brecciated[G] variety of tiger's eye known as pietersite found north-east of Outjo.

Zimbabwe is rich in deposits of agate, amethyst, aventurine, chalcedony, chrysoprase, citrine, jasper and milky, smoky and rose quartz. Noteworthy is an attractive green variety of chalcedony known as mtorolite, which is found in the Great Dyke at Mtorashanga.

Botswana has various deposits of crystal quartz, jasper and pink agates, the latter occurring in the Tuli region.

Swaziland has interesting deposits of chalcedony and rock crystal – some reach a length of 35 cm.

Mozambique has several deposits of agate along its southern border with Zimbabwe.

Quartz (carnelian). SA: tumbled stones

Serpentine

H 2-3

SG 2,3-2,6

Serpentine refers to a group of three minerals – lizardite, antigorite and clinochrysotile. Gemologists recognize serpentine as a species[G], mineralogists recognize it as a mineral group, while geologists speak of serpentinites – rocks composed of the serpentine minerals. Serpentine includes soapstone, talc and actinolite, which are used for carvings. It is found in metamorphic[G] rocks and is rich in magnesium.

Varieties and colours: Bowenite (translucent green, yellow-green); soapstone (grey-green); lizardite and antigorite (dark green).

Crystal system and mineral group: Trigonal and hexagonal silicates[G].

Serpentine. SA: 10,5 cm

Composition: Magnesium silicate[G].

Lustre: Earthy, greasy.

Identification points: Greasy, waxy feel. Softness: very easily fashioned and carved.

Size: Forms in veins[G] and massive lumps[G].

Status: Common.

Economic value: Ornamental carvings.

Distribution: Found in Mpumalanga and the Northern Province, South Africa. Grey-green soapstone found north-east of Windhoek in Namibia. A pale green variety occurs in the Nyanga area of Zimbabwe, which also has deposits of bowenite in the Mashava, Masvingo, Mberengwa and Kwekwe districts. Occurs north-west of Francistown, Botswana and in the north-east of the country.

Sodalite

H 5,5-6

SG 2,14-2,4

The easiest way to identify sodalite is by its dark blue colour. It is commercially exploited for carvings, stonework and jewellery. The largest southern African deposit of sodalite has been mined in northern Namibia.

Colours: Blue (light to dark), rarely white and yellow-green.

Crystal system and mineral group: Cubic silicate^G.

Composition: Sodium-aluminium silicate^G.

> **Did you know?**
> Sodalite is often confused with lapis lazuli, which is actually gem-quality lazurite.

Lustre: Greasy, vitreous.

Identification points: Distinctive dark-blue colour.

Size: forms as massive lumps^G.

Status: Rare.

Economic value: Lapidary^G material, used for *objets d'art*^{G} and ornamental stonework.

Distribution: Found in Kaokoland, Namibia and in the Mwami and Hurungwe areas in Zimbabwe.

Sodalite. Nam: 6 cm polished egg

Stichtite

H 1,5-2

SG 2,16

Often associated with chrysotile asbestos.

Colours: Purple, lilac.

> **Did you know?**
> Stichtite is one of very few purple minerals known. It is usually found in the serpentinites of ancient greenstone[G] belts as vein[G] fillings.

Crystal system and mineral group: Trigonal carbonate[G].

Composition: Magnesium-chrome carbonate[G].

Lustre: Waxy, greasy.

Identification points: Vibrant purple colour, softness – easily scratched with a fingernail.

Size: Forms as massive lumps[G] and vein[G] fillings.

Status: Very rare.

Economic value: Occasionally used for lapidary[G] purposes.

Distribution: Occurs in association with serpentine in several southern African greenstone[G] belts. At Kaapsehoop near Barberton, in the Murchison greenstone[G] belts, in South Africa, and in a few of the ancient greenstone[G] belts of Zimbabwe.

Stichtite (polished). SA: 8 cm

Sugilite

H 6-6,5

SG 2,74

Sugilite. SA: 8 cm piece

Sugilite created quite a stir when it was discovered in the manganese mines at Hotazel. It is known by several common names, including wesselite (after the Wessels mine where it originates) and royalazel (referring to the royal purple colour and the nearby town of Hotazel). It is very rare as a transparent crystal, which is in high demand.

Colour: Purple.

Crystal system and mineral group: Hexagonal silicate[G].

Composition: Potassium-aluminium-iron-lithium silicate[G].

Lustre: Earthy, vitreous (crystals).

Identification points: Distinctive purple colour.

Size: Usually found as massive lumps[G].

Status: Very rare.

Economic value: Lapidary[G] material, *objet d'art*[G] and, if transparent, as a gemstone.

Distribution: Found in the manganese mines, most notably the Wessels mine, north of Kuruman, in the Northern Cape, South Africa.

Titanite (Sphene)

H 5-5,5

SG 3,45-3,55

Titanite crystals are often twinned[G]. It occurs as a rare gemstone in Zimbabwe.

Colours: Yellow, green, olive-green, brown.

Crystal system and mineral group: Monoclinic silicate[G].

Composition: Calcium-titanium silicate[G].

Lustre: Adamantine, resinous.

Identification points: Wedge-shaped, twinned[G] crystals.

Size: Few mm to several cm (rare).

Status: Very rare.

Economic value: Rare gemstone.

Distribution: Found in the Northern Cape pegmatites[G] as well as the Richtersveld and the Pilgrim's Rest areas in South Africa. Attractive crystals up to several cm in the Gamsberg region, Namibia. Gem-quality yellow titanite from the Hurungwe district, in veins[G] of epidote at Mtoko and from the Matuki area, Zimbabwe.

Titanite. Zim: 2 cm

Topaz

H 8

SG 3,49-3,57

Topaz is a gemstone that has been known for millennia. The most popular colour is amber, the so-called imperial topaz from Russia. Excellent dark blue topaz has been mined in north-west Zimbabwe. Many kilograms of gem-quality colourless topaz have been mined from Namibia. Occurs in pegmatites[G] in granites.

Topaz. Zim: 10 cm

Economic value: Gemstone.

> **Did you know?**
> The name 'topaz' is thought to come from the Sanskrit 'tapas', meaning 'fire'.

Colours: Silver topaz (colourless); imperial topaz (yellow-orange), also white, grey, pink and blue.

Crystal system and mineral group: Orthorhombic silicate[G].

Composition: Aluminium silicate[G] with fluorine/hydroxyl.

Lustre: Vitreous.

Identification points: Hardness, excellent basal cleavage.

Size: Few mm to several cm.

Status: Rare.

Distribution: Topaz is found in a few pegmatites[G] in the Northern Cape, South Africa. Silver topaz from Damaraland, Namibia. Other varieties from the Karibib pegmatites[G], the Brandberg and the Omaruru districts. In Zimbabwe, beautiful blue topaz and colourless topaz occur in the Mwami area. In Swaziland, topaz deposits are found in pegmatites[G] in the foothills of the Sinceni Mountains.

Tourmaline

H 7

SG 3,03-3,10

Tourmaline is popular in jewellery because it can form absolutely clear crystals with beautiful colours. Found in pegmatites[G].

> **Did you know?**
> Single crystals can exhibit multiple colour variations, such as the watermelon tourmaline.

Varieties and colours: Elbaite, including indicolite (blue-green); rubellite (pink); schorl (black); dravite (brown); red, green, blue, pink, peach, green, white and colourless.

Crystal system and mineral group: Trigonal silicate[G].

Composition: Complex sodium-aluminium borosilicate.

Lustre: Vitreous.

Identification points: Elongate pencil-like crystals; crystal surfaces often have multiple, parallel striations[G].

Size: Few mm to tens of cm.

Status: Rare.

Economic value: Gemstone and lapidary[G] material.

Distribution: Found in South Africa in pegmatites[G] in the Northern Cape as well as in the Kenhardt district. In the Karibib, Namibia, excellent quality rubellite and indicolite is exploited, while indicolite (blue elbaite) occurs at Neu Schwaben. In Zimbabwe, green tourmaline occurs at St Ann's mine and rubellite in the Karoi area. Rubellite occurs near Kubuta in Swaziland.

Tourmaline (watermelon). Nam: 2,6 cm

Verdite

H variable*

SG variable*

Verdite is unique – several other green rocks that are often referred to as verdite are usually another variety such as buddstone, which contains folded layers of white chert^G.

Verdite. SA: 8,7 cm

Colours: Green, with yellow-gold flecks.

Crystal system and mineral group: Combination of minerals (rock).

Composition: Mixture of green minerals.

Lustre: Earthy, dull.

Identification points: Distinctive dark green rocks with light green interlayered portions.

Size: Forms as massive lumps^G.

Status: Very rare.

Economic value: Lapidary^G material, animal carvings.

> * Verdite is a rock made up of several minerals. Therefore it does not have its own **SG** or **H** value.

Distribution: Barberton and Sabie districts in South Africa; and in the Concession area, Zimbabwe.

53

Glossary

Alluvial: Process whereby detritus (loose clay, sand, gravel) is transported and deposited by a river.

Amorphous: Describes minerals that do not have a characteristic external shape as they lack crystalline structure.

Asterism: An optical effect caused by light reflections within the stone, which impart a star-shaped pattern to the gemstone. When the stone is rotated, the lines appear to move across the surface.

Biotite: A member of the mica group of minerals that is dark brown to dark green in colour.

Borate: Mineral containing the chemical elements borate and oxygen as BO_3.

Brecciated: A 'fragmental' rock made up of many pieces of coarse, angular, jagged pieces of broken rock.

Carbonate: Referring to a mineral that contains the anionic structure CO_3^{-2}, for example, calcite, $CaCO_3$. All carbonates react in acid by fizzing.

Chert: Microscopic hydrated silica, either inorganic or organic in origin.

Conchoidal: Describes the curved, shell-like surface form of a brittle substance. A fracture pattern that resembles broken glass – many curved surfaces.

Dichroism: The ability of a crystal to show two different colours.

Dodecahedron: Crystal form with 12 sides.

Eluvial: A type of soil formed by the natural weathering of rock.

***en cabochon*:** A gemstone that is cut and polished into a shape with a flat base and a polished dome, with no facets.

Gneiss: A layered, coarse-grained metamorphic rock made up of relatively large minerals.

Greenstone: Rocks coloured green by secondary minerals such as chlorite, fuchsite, actinolite, and epidote.

Hydroglossular: Mineral of the hydrogarnet group.

Igneous: A rock or mineral that has solidified from molten magma.

Kimberlite: A volcanic rock containing olivine-magnesium and iron-bearing minerals. A host rock for diamonds.

Lapidary: The art of cutting and polishing gemstones.

Lump: Non descript mass of material, e.g. a lump of coal.

Metamorphic: Rocks altered by the action of pressure and/or heat.

Objet d'art: A small object considered to be of artistic value, e.g. Fabergé eggs.

Octahedron: Crystal form with 8 crystal faces.

Oxide: Mineral/ gemstone containing oxygen linked to other elements.

Pegmatite: A coarse-grained igneous rock that forms in veins or lenses. Usually the last to crystallize from magma. Can contain large amounts of rare elements and minerals.

Pleochroism: Ability of a mineral to show different colours due to differential absorption of various wavelengths of transmitted light.

Pseudomorph: or 'false form'. A mineral whose crystal appearance is like that of another. Forms by chemical alteration, replacement, encrustation or substitution.

Replacement mineral: A mineral that chemically replaces another (see 'pseudomorph' above).

Schist: A metamorphic rock that is made up of thin layers.

Sedimentary: Rocks formed from the accumulation of detrital material that is deposited by wind, water or ice.

Silicate: Mineral with a crystal structure containing SiO_4 tetrahedra.

Sixling: Twinned crystal with 6 radiating arms from centre point – like a snowflake.

Species: Gemstone\mineral distinguished from others by unique chemical and physical make-up; some species have varieties.

Striations: Series of parallel lines or grooves etched on crystal face of gemstones and minerals.

Twin: An ordered intergrowth of two or more single crystals.

Vein: A gash in a rock filled with secondary minerals.

Credits and Further reading

Specimen credits

Pictures are listed with page number from left to right and top to bottom using the letters a,b,c etc. All photographs and specimens are by Bruce Cairncross, except those specimens mentioned below.

Eric Farquharson	= EF
Rob Smith	= RS
Desmond Sacco	= DS
Martha Rossow	= MR
MuseumAfrica	= MA
Transvaal Museum	= TM
Nollie Cloete	= NC
De Beers	= DB
Theresa Cairncross	= TC
Rand Afrikaans University	= RAU

Pages:

Cover b - DS, cover c - EF , 1b - DS,
3 a&b - DS, 5a - RS, 5b - NC,
6a - RS, 7a - DS, 7b - MA,
18a - DS, 18b - DB, 18d - RS,
18f - TC, 19d - MR, 19e - DS,
21a - DS, 21b - RS, 21c - EF,
23 - EF, 24 - RS, 25 - DS, 26 - TM,
29 - RAU, 30 - EF, 35a - DS,
36 - DS, 37 - EF, 38a - RS, 39 - EF,
41 - RS, 42 - EF, 43a - EF, 43b - EF,
44a - EF, 44b - EF, 45 - EF, 47 - EF,
49 - RS, 51 - RS, 52 - EF, Back cover a - RS, back cover b - EF

Further reading

Cairncross, B and Dixon, R. 1995. *Minerals of South Africa.* Geological Society of South Africa: Johannesburg.

Cipriani, C. 1986. *The Macdonald Encyclopedia of Precious Stones.* Macdonald & Co. Publishers: London.

Macintosh, EK. 1983. *Rocks, Minerals and Gemstones of Southern Africa – a Collector's Guide.* Struik (Pty) Ltd: Cape Town.

McIver, JR. 1966. *Gems, Minerals and Rocks in Southern Africa.* Purnell and Sons (SA) Pty Ltd: Johannesburg.

Mineral Resources of Namibia various authors 1992. Ministry of Mines and Energy, Geological Survey of Namibia, Windhoek: Namibia.

Wilson, MGC and Anhausser, CR. (1998). *The Mineral Resources c South Africa.* Council for Geoscience Handbook 16, Pretor.